Emergency Procedures for Employees with Disabilities in Office Occupancies

Federal Emergency Management Agency

United States Fire Administration

C O N T E N T S

With funding from the United States Fire Administration, this guide was develop by the National Institute of Standards and Technology with assistance from the National Task Force on Life Safety and People with Disabilities.

For additional copies of this publication write to:
United States Fire Administration
16825 South Seton Avenue
Emmitsburg, Maryland 21727

Review Panel

The following individuals comprised our Review Panel. Their willingness to serve through the planning session, resource, draft review.... sometimes at a moment's notice and always ready with advice and information for inclusion in this guide.

Brian Black
Eastern Paralyzed Veterans Association

Marianne Cashatt
Consultant. Disability Awareness/Public Relations

Alan Clive
Federal Emergency Management Agency

Eunice Fiorito
Rehabilitation Services Administration

Victor Galloway
Rehabilitation Services Administration

Anne Hirsh
Job Accommodation Network, affiliate of *PCEPWD*

Marsha Mazz
U.S. Architectural 6 Transportation Barriers Compliance Board

Bill Scott
Abilities Unlimited

Certain commercial equipment or products are identified in this guide as representative examples of products which are available for the purposes discussed. Such identification does not imply endorsement by the U.S. Government nor does it imply that the equipment identified is necessarily the best available for the purpose.

People with disabilities are increasingly moving into the mainstream of society, contributing to the diversity which has been this country's strength. It is only right that they be provided with the same level of safety as the rest of society, as referenced in the Americans with Disabilities Act (ADA). Equipment and procedures exist that can provide such safety for any person with a disability that is not so severe that it would preclude the ability to work. The key points regarding finding the best solution for your building are, first, to remember that every person with a disability has unique abilities and limitations, and accommodations should be tailored to their needs. Second, it is crucial that the person be included in the decision on which equipment and procedures will work for them to provide them with the confidence that they will be protected.

It is every employer's responsibility to provide a safe place for all employees to work. Employees with disabilities are entitled to THE SAME level of safety as everyone else (no more/no less). The "reasonable accommodation" as mandated in ADA is intended only to provide this same level of safety and utility as is provided to everyone. Further, we cannot predict when any one of us may need assistance, such as in the case of a broken leg or the development of heart disease.

The underlying principle in providing safety from fire and smoke in buildings is that of safe egress: the efficient relocation of building occupants to an area of safety usually outside the building. This depends on several steps. First, we must provide for **detection** of a fire, before it can interfere with the movement of people. Next comes **notification** of the people that a potential danger exists and that evacuation to a predetermined point of safety should begin. Third is the **movement** of people through the building spaces to a protected exitway by which they can leave the building.

The techniques for *detection, notification* and *movement* are generally appropriate for anyone in any setting, but there are some exceptions. For example, special considerations are required with regard to *movement* in the case of the "limited mobility" of the patients in health care occupancies and the "limitations imposed" on the occupants in correctional occupancies. Both of these occupancies require special considerations with regard to movement, or a higher

level of protection (such as "defend in place"), which does not require movement. For other occupancies, the presence of individuals with temporary or permanent disabilities requires some additional planning.

The purpose of this guide is to provide information for facilities managers and may be useful for those individuals who might need special assistance as to the notification of an emergency situation and/or in the evacuation of a building. The information includes examples of equipment available as well as suggestions on procedures and comments on some of the advantages and the disadvantages. By starting with the same information, options can be discussed and a decision on the best approach to providing for the individuals needs can be made. This discus-

sion is a crucial step because each person's capabilities and limitations are unique; thus plans must be designed to meet the needs of the individual to be most effective.

Detection

The detection of fires is generally accomplished by automatic systems that do not require human intervention. Generally, no special accommodations are needed for people with disabilities.

One exception is where manual pull stations are provided in public buildings by which persons can initiate a fire alarm if they discover a fire before it is detected by the automatic system. In recent years, codes have been revised to require that these manual pull stations be mounted at a height to be within the "reach range" of 48" to 54" for the person in a wheel chair. Facility managers

Alarms: Stroke and Horn, *left,* and Stroke for the Hearing Impaired, *right,* are "ADA" and "UL Standard 1971" compliant.

should also consider that not all people possess the strength and/or dexterity to operate some of the manual pull station devices (e.g., those with arthritis or quadriplegia).

Notification

Notification refers to the process of informing occupants that an emergency exists and that some action is needed. In most cases, this action is simply to evacuate, and the quantity of information to be given is only this fact. Traditionally, notification of an emergency has been accomplished by audible devices, which are effective

for all but those with hearing impairments. Recently, visible devices (high intensity flashing lights) are being used along with the audible devices to broaden the range of notification effectiveness.

In larger buildings, emergency evacuation may involve relocation to a safe area within the building or sequencing evacuation by floor or area so as not to overload the stairways. In such cases, the amount of information that must be provided to occupants is substantially greater. This is typically done audibly, through emergency paging systems. These are effective for all except those with hearing impairments where textural displays, (television monitors or scrolling text signs) are located throughout the building, or portable devices (tactile/vibratory pagers) have been utilized effectively.

Movement

By far, the greatest range of special needs exists in the area of movement of persons to safe areas. People using wheelchairs or with other obvious mobility disabilities come immediately to mind; but, there are many who may not appear to have a disability who will also require some special assistance.

Permanent conditions such as arthritis or temporary conditions such as a sprained ankle or a broken leg can limit one's ability to evacuate quickly and safely. Heart disease, emphysema, asthma, or pregnancy can reduce stamina to the point of needing assistance when moving down many flights of stairs.

One major challenge is the identification of those individuals who may need the special assistance. Consider persons with emphysema, asthma and other respiratory conditions who may perform well in a drill but then experience problems in an actual emergency situation, as learned in the World Trade Center evacuation as a result of the February 1993 bombing. The people with respiratory conditions who were interviewed, described the terror they experienced when faced with the grim reality of extreme exertion required to escape down the many flights of stairs in unfamiliar and smoke-filled stair towers. They also explained that prior to that emergency evacuation they had never considered themselves as having a disability that would qualify them as potential candidates for inclusion in the emergency evacuation plans for those requiring special assistance.

How to Proceed

There will always be someone who will need some special assistance in the event of a fire or other emergency requiring evacuation. Thus, identifying these individuals is essential, never losing sight of the fact that some of these people may not recognize their own need for assistance. In addition, allowances for visitors present in the building must also be made.

Once identified, individuals should be consulted about their specific limitations and how best to provide assistance. Finally, the methods for accommodation and assistive devices should be selected and discussed. This is necessary to assure a safe "emergency" evacuation from the building for the individual with a disability.

The remainder of this guide is intended to present the range of possible approaches to the accommodation of special needs along with the advantages and disadvantages. From this, it is expected that the facility management and the affected individuals can make a more informed decision on what approach would work best for the specific conditions present. All concerned should operate with a common understanding of the options so that an optimum solution can be reached.

Identifying Special Needs

While the Americans with Disabilities Act of 1990 (ADA) does not require formal emergency plans, Titles I and III do require that policies and procedures of public accommodations be modified to include people with disabilities. The facility emergency plan which may already have provisions for individuals with limited mobility must now include all the other classifications of disabilities as covered in the ADA. These include:

■ Individuals with varying degrees of mobility impairments, ranging from slow walkers to wheelchair users.

■ Individuals who are visually impaired and may require special assistance in learning the emergency evacuation routes or assistance in proceeding down exit stairs.

Intele-Modem converts a personal computer into a TDD. It also automatically converts PC (ASCII) to TDD (Baudot); directly connects to phone lines; automatically detects both Baudot and ASCII calls; and is FCC approved.

Clarity Phone with built-in equalizer automatically tunes, tones and balances sounds to improve clarity, making words clearer, not just louder. The phone is designed to help those with high frequency hearing loss, a problem affecting 95% of the people who are hard-of-hearing. The Clarity adjustment controls set the high frequency that is right for each individual.

■ Individuals with hearing impairments and who may require modification to the standard audible alarms.

■ Individuals with temporary impairments due to recovery from serious medical conditions such as stroke or traumatic injuries such as a broken leg or a sprained ankle or surgeries such as a knee or hip replacement.

■ Individuals with medical conditions such as respiratory disorders or pregnancy who may tire easily, need special assistance or more time to evacuate.

■ Individuals with mental impairments who may become confused when challenged with the unusual activity during an emergency, lose their sense of direction, or may require having emergency directions that are broken down into simplified steps or basic concepts.

■ Other populations that need to be considered as being vulnerable, such as visitors or customers with small children who require extra time to evacuate down stairs, or employees who work outside the normal working hours. All of these individuals need to have special provisions or contingencies included in the emergency plan for their protection.

Discussion with the Individual

Keep in mind that someone with a permanent or major impairment generally knows the best way to be assisted. A minute or so spent talking with the individual will give you crucial information. People providing assistance should be trained on how to help without causing injury to themselves or others. This is especially relevant if someone needs to be lifted or carried.

Put in Writing

Identify and plan for times (of the day and the week) plus locations in the workplace where the basic life safety or emergency contingency plans have not been put in place or due, to some other factor, might not work.

Periodic Review

Innovative educational techniques such as role playing or the use of audiovisual aids might prove more effective than more traditional methods of information dissemination used in the past. Practice using the elements of sections II and III that you have selected as being appropriate for your workplace. Practice will instill confidence in one's ability to cope in an

Shake-Up System. 9-V battery smoke detector, signal unit, and vibrator alert device. Detector and receiver attach to wall and/or ceiling with a vibrator device placed on the desk. A smoke detector sends a transmitted signal to the receiver and activates the vibrator.

emergency. It will also do more than anything else to assure that appropriate lifesaving actions will be taken during a real emergency.

Practice consists of one of three types of activity, - walk through procedures, announced drills or surprise drills.

Walk through Procedures

Practice separate parts of a plan one at a time. In this way you can concentrate your efforts on the particular- parts and par.-

ticular individuals requiring more extensive practice. Members of an emergency response organization (e.g.,

fire wardens) would be prime candidates for this practice This is also a way of introducing newly hired employees in the workplace to important parts of the plan.

Announced Drills

As with the walk through procedures, this is intended more to train than to evaluate. Such drills will help identify crucial coordination activities and communication links.

Surprise Drills

Use these drills infrequently. Depending on the situation, this might be done once or twice each year. Surprise drills should involve some realistic elements (e.g., blocked exits.)

Panic has rarely been reported, either in drills or in actual emergencies.

Ultratec Superprint ES This portable, 20-character display Telecommunications Device for the Deaf (TDD) provides printed records of conversations and optional auto-answer.

Notification Appliances

The disability that most affects the process of notification of an emergency condition is hearing impairment. Hearing impairments can range from mild hearing loss to an extreme of profound deafness, the level at which individuals receive no benefit from aural input.

Many persons who are hearing impaired can use their residual hearing effectively with assistance from hearing aids or other sound amplification devices, often augmented by lip reading.

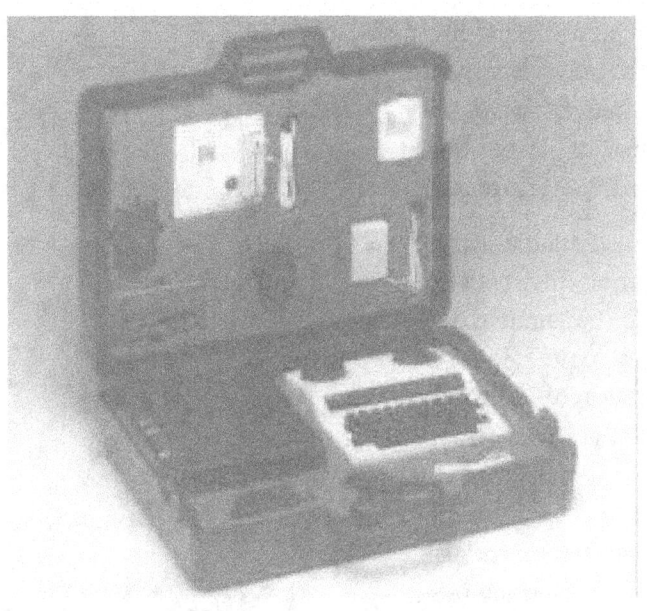

All-in-One Carrying Case includes: TDD; telephone signaler; telephone amplifier; door knock signaler; telecaption decoder; notification vibrator; and visual/audio smoke detector.

However, hearing aids also amplify background sounds, and the sound of the emergency alarms may interfere with or even drown out voice announcements of an emergency, voice communication system.

Systems used for emergency notification must comply with UL1971, the Underwriters Laboratories Standard for Emergency *Signaling* Devices *for the Hearing Impaired*. The signaling devices covered in the UL 1971 standard are designed to alert persons with hearing impairments through the use' of light, vibrations, and air movement.

Many hotels post a sign at the desk to make deaf or hearing impaired guests aware that rooms with strobe lights arc available.

In some government buildings, employees who are deaf or hard of hearing have been provided with tactile/vibratory pagers to notify them when a fire alarm has been activated.

Building managers who wish to provide wheelchair assist equipment for use by visitors can receive assistance in selecting appropriate devices from one of the groups in the Resources section on page 25.

This device was used in the study conducted in 1988 by Underwriters Laboratories to evaluate alternative signaling systems that would alert persons with hearing impairments to fires and other emergency conditions. The study established guidelines for the use and installation of the devices covered in "Standard 1971."

Tactile Signage-Raised Text and Braille

Braille signs have been installed at some locations in buildings to assist individuals with visual impairments. You may have noticed these raised patterns of dots on elevator control panels. The problem with the use of such labels to mark egress doors is that the person must be at the door in order to feel the label. Thus, they provide no directional guidance on how to find the door in the first place.

Audible Directional Signage

Audible remote signage is a way of informing individuals who are visually impaired of what they need to know about their environment.

Audible directional instructions are transmitted by low power radio waves or infrared beams. The signal/ instructions are then picked up by a small receiver carried by the individual (e.g., "the exit is 25 steps south of the front desk," or simply "stairway," "restroom," or "elevator") act as signals when one approaches a stairway, rest-room or elevator.

Audible Pedestrian Systems

Another example of audible signs

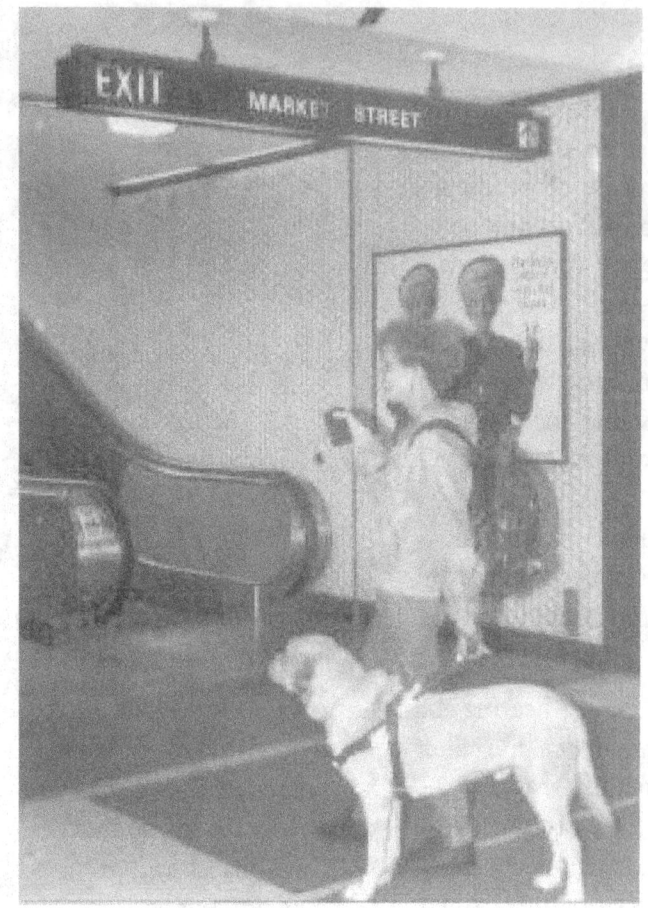

Talking Signs ™ provide people who are blind with the directional and usage clues that traditional visual signs provide for sighted persons. By sending information from installed infrared transmitters these signs speak for themselves. Hand held sensors pick up information from the transmitters and give verbal directions to the blind individual (see illustration above). These signs are installed in the San Francisco Municipal Railway and Bay Area Rapid Transit District.

is the common pedestrian traffic signal. These signs "cuckoo" and "chirp" to alert pedestrians to changing traffic signals. But, these devices have some inherent limitations for those with learning disabilities. They are not currently in use for emergency egress systems within buildings, although there are exit signs available that flash and sound internal horns when activated by the building fire alarm system.

Movement Aids/Equipment

Another area where disabilities impact on emergency egress is with mobility limitations. This is most frequently associated with wheelchair users. Here we should be sensitive to the fact that wheelchairs represent mobility and are frequently fitted to accommodate the specific physical needs of the user.

Thus, whether evacuated with or without their wheelchairs, they will need their own chairs when they reach safety for both physical and psychological reasons.

Permanently Installed Systems

There are several types of controlled descent devices that can be permanently installed within stairways to accommodate wheelchair users. In some, the individual transfers from the wheelchair to the portable, controlled descent chair. Some models permit a relatively small person to transport a larger person while with other devices, the individuals ideally should be about the same weight. These chairs are designed to travel down stairs on special tracks with friction braking systems, rollers or other devices to control the speed of descent.

Another type of controlled descent device is designed so the wheelchair user rolls onto the transport device and the wheelchair is secured to the device. This has the advantage that the wheelchair user does not have to be separated from the chair - a situation that will be more comfortable and reassuring.

The wheelchair lift is a motor-driven device designed to be installed in a stairway. Vertical wheel-

Wheelchair lifts for use indoors,top, and for use outdoors, bottom.

Garaventa **EVACU-TRAC** [TM]
Developed in Switzerland.
Top: Convenient, top of
stair storage. *Middle:*
1. Brake system engages
when lever is released;
2. Adjustable safety belts;
3. Rubber tracks grip
stairs; 4 Eight auxiliary
wheels for smoother ride
on flat surfaces, such as
stair landings. *Bottom:*
Designed so a passenger's
weight propels it down
stairs. Governor limits the
maximum descent speed.

EVAC+CHAIR [TM] 300-H'
Top: Folds for on-the-job
storage. Can be readily
available for emergencies
Middle: Unfolds/opens
quickly and Weighs only
15 pounds but has a 300 lb.
capacity. Cantilevered
design places seat inches
above stairs. Other fea-
tures: sliding head rest;
quick-release safety belt
buckle; and instructions
permanently stamped on
back. Bottom: Changes the
obstacle of firestairs into
usable escape route for
all, e.g., pregnant women,
the frail or employees with
limited stamina, or some-
one with a temporary
disability.

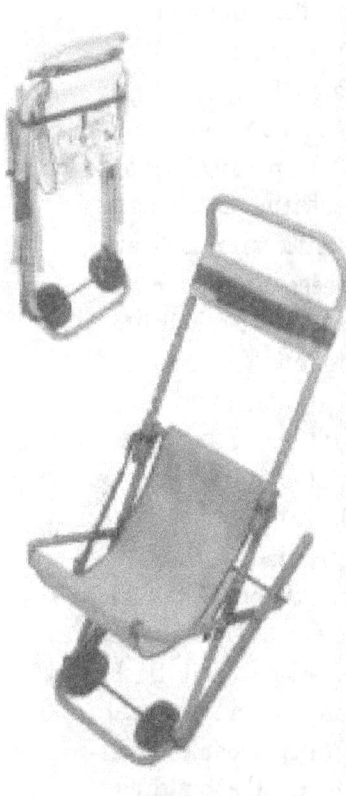

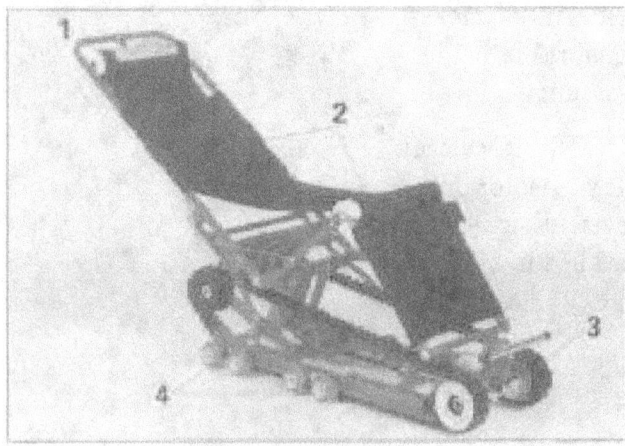

Scalamobil ™

Stairclimbing and power unit, invented in Germany. Three-step process for use. Top: First, attach handles to the Scalamobil. Second, attach Scalamobil to wheelchair. Third, begin operation. Middle: 12V/12AH power base. Bottom: Operator's safety features include: automatic mechanical security brakes on every wheel; variable speed control from 6-12 steps per minute, and ability to park the wheelchair safely on any step during ascent or descent. Designed to negotiate most all stairs, from the extremely narrow stair to curving circular stairs.

Chair lifts are differentiated from elevators in that they are' limited in the height of their vertical lift, are not enclosed, and do not go through a floor level. These

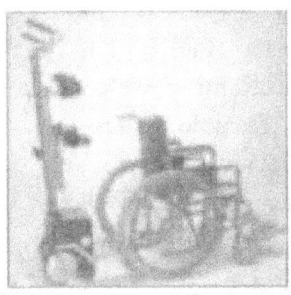

lifts were originally intended for private residences, but are now being used in nursing homes, churches and public buildings.

Always consult

the wheelchair user as to the selection of an emergency evacuation chair. The advantages or disadvantages of these devices are dependent on the capabili-

Evacuation Assistance Device

A three-person, assisted-wheelchair-carry device, called "Evac-u-Straps," was developed by a wheelchair user. It consists of wide padded leather wrist bands with velcro closures equipped with large metal grasping hooks. The hooks are designed to be attached to both sides of the front of the wheelchair. Persons on either side of the wheelchair grasp the straps and are assisted by a third person behind, keeping the wheelchair slightly tipped backwards. The wheelchair user assists by hand-braking the wheels.

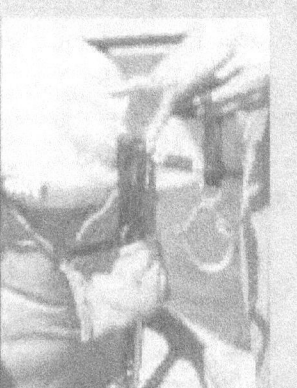

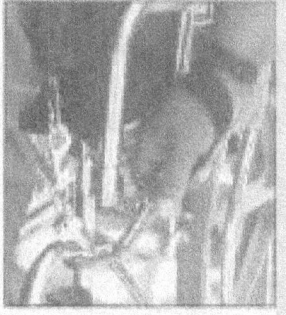

Photos by John Pauls, Building Use and Safety Institute

ties, acceptance, and understanding of the end user(s). The effectiveness or failure of evacuation chairs as a rule can be attributed to the fact that the wheelchair user was not consulted as to the equipment selection. Chairs that do not accommodate the physical needs of the user create problems which may lead to a refusal to use them in an emergency.

Elevators

Most people are familiar with the fact that elevators are not to be used for emergency egress and are so marked in most buildings. Elevator codes require that when smoke detectors in elevator lobbies activate, the elevator is recalled to the ground floor (as long as the ground floor smoke detector is not the one that alarmed) and is taken out of service. The fire department can operate the elevator with a special key and may use it to move their people

and equipment, or for evacuation of occupants. This means that without the fire department, persons with disabilities are relegated to the stairs or must await rescue.

In recent years (especially since the 1993 World Trade Center bombing), there has been a growing interest in providing elevators that can be used for emergency evacuation. In a study conducted for the General Services Administration (GSA), the National Institute of Standards and Technology (NIST) found that the use of both elevators and stairs can improve evacuation times by as much as 50% over stairs alone.

However, elevators that are used for emergency evacuation need to be specially designed to assure their reliability and safety during the fire. NIST research has shown that, with enclosed lobbies at each floor

which are pressurized through the shaft so that both remain smoke free, dual power systems for reliability, and water resistant components to prevent failure due to flooding of the shaft by firefighting water, it is feasible to design elevators that are sufficiently safe to allow their continued use for emergency evacuation. (Feasibility of Fire Evacuation by Elevators at FAA Control Towers, NISTIR 5445, 1994.)

Miscellaneous Devises

A number of unique escape devices have been developed over the years. These include controlled descent devices using cables and chutes of various types. The cable devices usually use a strap or chair secured to the cable by a device that is squeezed to allow descent. The more you squeeze, the faster you go. Letting go stops your descent. Most peo-

ple are reluctant to evacuate down the outside of a building.

The chutes may be solid or flexible labric tubes that generally rely on friction to control speed. They have the advantage that they don't let you see out, so they are more acceptable than cable devices. However, their acceptance in practice in this country has been limited.

There is little information available as to the performance of these devices in emergency situations. These unique specialized escape devices generally have serious shortcomings. (Egress Procedures Technologies for People with Disabilities. Final Report of a State of the Art Review with Recommendations for Action, ATBCB 1988.)

Sprinklered Buildings

In a study of areas of refuge conducted by NIST for the GSA, it was concluded that the operation of a properly designed and maintained sprinkler system eliminates the life threat to building occupants regardless of their individual abilities and can provide superior protection for people with disabilities. Sprinkler systems will, in most circumstances, provide the protection to permit evacuation that is limited to the area under immediate threat from the fire. In sprinklered buildings it would probably be appropriate to put more emphasis on understanding of the protection afforded or provided with the sprinklers and about limited evacuation through horizontal exits versus total evacuation from the building.

Of course, while about 95% reliable, there is a small possibility that the sprinkler system will fail to extinguish the fire. For these possibilities, there need to be contingency plans for providing evacuation assistance for all occupants, including those needing special assistance.

Area of Refuge/ Rescue Assistance

Even in buildings equipped with sprinkler systems it is recommended that areas of refuge be provided. There is the small possibility that the sprinkler system will fail to extinguish the fire and there is the problem of smoke propagation. It is quite possible for a person with a disability to be stranded and overcome with smoke before the arrival of the rescue personnel, given the difficulty in locating someone in a smoke-filled building. For these possibilities, there need to be contingency plans for providing evacuation assistance for all occupants, as well as those needing special assistance.

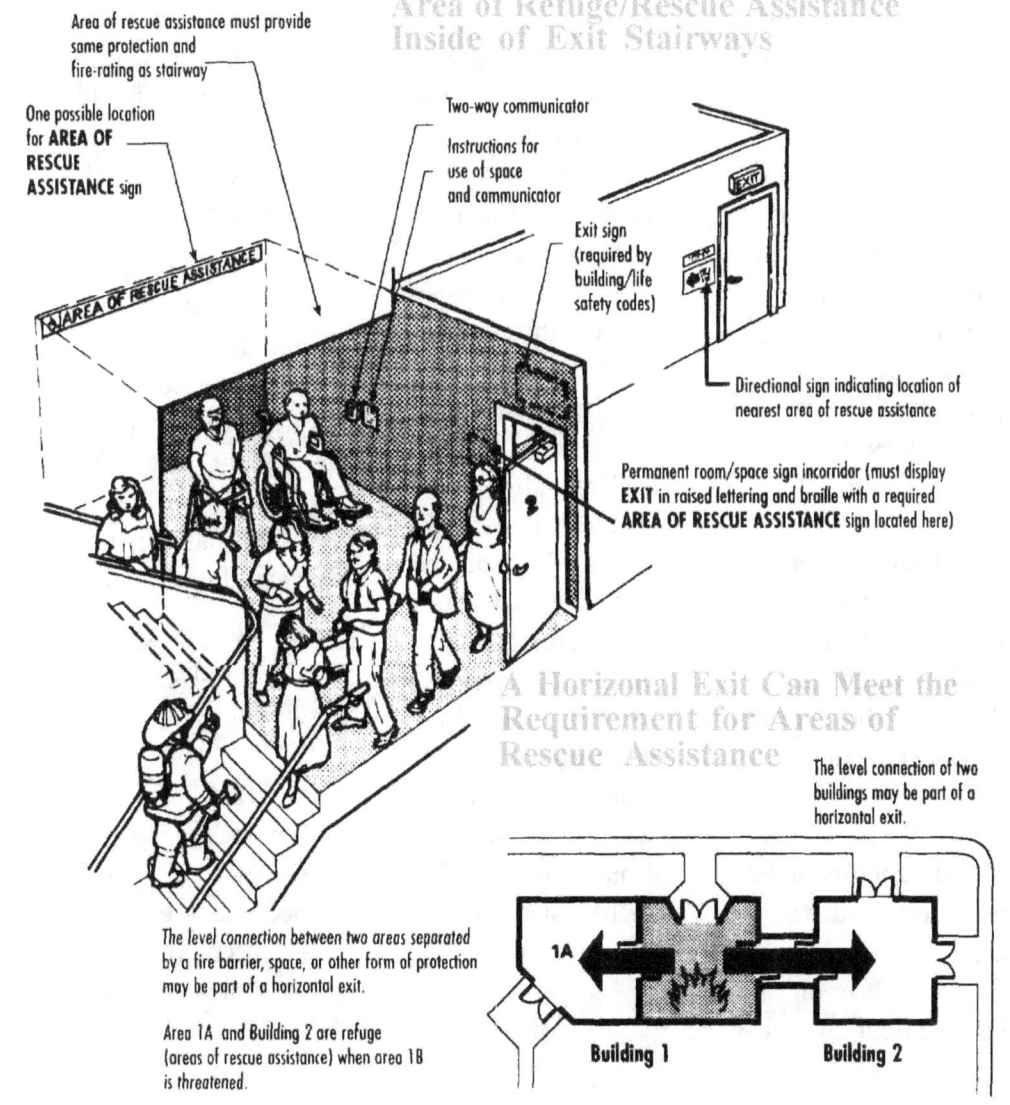

Area of rescue assistance must provide same protection and fire-rating as stairway

One possible location for **AREA OF RESCUE ASSISTANCE** sign

Two-way communicator

Instructions for use of space and communicator

Exit sign (required by building/life safety codes)

Directional sign indicating location of nearest area of rescue assistance

Permanent room/space sign in corridor (must display **EXIT** in raised lettering and braille with a required **AREA OF RESCUE ASSISTANCE** sign located here)

Area of Refuge/Rescue Assistance Inside of Exit Stairways

A Horizonal Exit Can Meet the Requirement for Areas of Rescue Assistance

The level connection of two buildings may be part of a horizontal exit.

The level connection between two areas separated by a fire barrier, space, or other form of protection may be part of a horizontal exit.

Area 1A and Building 2 are refuge (areas of rescue assistance) when area 1B is threatened.

1A

Building 1   Building 2

Identifying Those with Special Needs

Before special accommodations can be made, persons needing them must be identified. One strategy is to maintain a listing of individuals needing assistance and keeping it current as part of the facility's emergency plan. At the beginning of employment during the orientation process is the time to begin to stress the importance of identifying if an individual will need special assistance. Of course, since conditions change and persons can become temporarily disabled, this system needs to be flexible.

Such lists must be accessible by the emergency personnel to assist in the emergency evacuation. But, it should be understood that there are many individuals who are protective of their right to independence and privacy and who may be reluctant to have their names put on such a list. Some disability categories are easily recognizable and in these cases the individual can be approached as to what can be done to assist them in emergency evacuation.

It is important to treat the individual as one who happens to have a particular disability, and not make the mistake of "lumping" together all persons with disabilities in the development of emergency procedures. There are some emergency plans (and codes on which they are based) where all persons with disabilities were "directed" to go to the area of rescue assistance to await members of the emergency team to escort them to safety. As a general rule there is no reason that individuals who are blind or deaf cannot use the stairs to make an independent escape as long as they can effectively be notified of the need to evacuate and can find the stairway).

One of the lessons learned from interviews of people with disabilities following the February 1993 World Trade Center bombing was that, prior to the incident, some of the people with disabilities said that, in the interest of privacy or because they felt that they did not need special assistance, they) had opted not to identity) themselves to be among those listed as disabled in the emergency management plan. They realized after the incident that they did need assistance and that they had not realized how vulnerable they were outside of normal working hours when there were few co-workers around to provide such assistance.

"Buddy" Systems and Fire Wardens

Buddy systems are widely accepted and used, but have some inherent faults or flaws. When setting up such a system in the workplace, consider the following potential problem areas and potential solutions.

To be effective, the person and the buddy must be able to make contact with each other quickly when the need arises. Situations that can prevent this include:

- The "buddy" is in the building, but is absent from the customary work area
- The "buddy" cannot locate the person with a disability because the person is absent from the customary work area

The officials coordinate immediate emergency response, call the employee and alert responding fire service.

- Employees with disabilities can be given the responsibility for selecting their own "buddies"; bimonthly emergency plan reviews should include checking the status of "buddies."

- The "buddy" is trained by the employee with a disability as soon as they are recruited.

- The employee with a disability is encouraged to select only "buddies" who are capable. Practice sessions are required to ensure that "buddies" can handle their assigned tasks.

- Employees with disabilities are encouraged to select only friends/colleagues as "buddies."

- The employee with a disability is working late, etc., when the "buddy" is unavailable

- The "buddy" has left the company) and a new one has yet to be identified

- The "buddy" has not been trained in what to do or how to assist

- The "buddy" is inappropriate (e.g., not strong enough)

- The "buddy" isn't acceptable to the employee with a disability

- The "buddy" forgets or is frightened and abandons the employee with a disability

Now, consider the lollowing potential solutions:

- Assign at least "two buddies" who are work associates. Alert the floor warden about the work location of the person with a disability.

- If he/she cannot locate the assigned person, the "buddy" should alert the floor warden. Employees could be given pagers.

- Employees with disabilities should identify themselves to the officials in the emergency control center when in the building after hours.

New York City leads the nation in a number of techniques for addressing fire safety in tall buildings, including the designation of fire wardens. Under Local Law 5, a fire warden is assigned for each floor of a building, and is responsible for the safe evacuation of persons on that floor. The fire warden knows who is and is not at work that day, what visitors are present, and who might need assistance in case of emergency. New York fire wardens take required training at regular intervals. The law also requires a building fire safety manager whose full-time job is to keep fire emergency plans up to date and who coordinates the activities of the fire wardens with the fire service during an emergency.

Vision Impairments

When assisting persons with vision impairments there are some basic. rules to follow in order to be effective.

■ Announce your presence; speak out when entering the work area.

■ Speak naturally and directly to the individual and NOT through a third party. Do not shout.

■ Don't be afraid to use words like "see," "look," or "blind."

■ Offer assistance but let the person explain what help is needed.

■ Describe the action to be taken in advance.

■ Let the individual grasp your arm or shoulder lightly, for guidance. He/she may choose to walk slightly behind you to gauge your body reactions to obstacles; be sure to mention, stairs, doorways, narrow passages, ramps, etc.

■ When guiding to a seat, place the person's hand on the back of the chair.

■ It leading several individuals with visual impairments at the same time, ask them to hold each other's hands.

Suggestions When Assisting Owners of Dog Guides

■ Do not pet or offer the dog food without the permission of the owner.

■ When the dog is wearing its harness, he is on duty; if you want the dog not to guide its owner, have the person remove the dog's harness.

■ Plan for the dog to be evacuated with the owner.

■ In the event you are asked to take the dog while assisting the individual, it is recommended that you (the helper) hold the leash and not the dog's harness.

■ You should ensure that after exiting the building that individuals with impaired vision are not abandoned" but are led to a place of safety, where a colleague(s) should remain with them until the emergency) is over. Another of the lessons learned from the World Trade Center incident involved the complaints of blind tenants who after being escorted down and out of the building, were unceremoniously left in the unfamiliar environs out-of-doors in the midst of a winter ice storm, where they had to negotiate ice covered sidewalks and falling glass from overhead.

Hearing Impairments

When assisting persons with hearing impairments there are also some things to keep in mind. These include:

■ Flick the lights when entering the work area to get the person's attention.

■ Establish eye contact with the individual, even if an interpreter is present.

■ Face the light, do not cover or turn your face away, and never chew gum.

- Use facial expressions and hand gestures as visual cues.
- Check to see if you have been understood and repeat if necessary.
- Offer pencil and paper. Write slowly and let the individual read as you write. Written communication may be especially important if you are unable to understand the individual's speech.
- Do not allow others to interrupt or joke with you while conveying the emergency information.
- Be patient, the individual may have difficulty comprehending the urgency of your message.
- Provide the individual with a flashlight for signaling their location in the event that they are separated from the rescuing team or buddy and to facilitate lip-reading in the dark.

Learning Disabilities

Persons with learning disabilities may have difficulty in recognizing or being motivated to act in an emergency by untrained rescuers. They may also have difficulty in responding to instructions which involve more than a small number of simple actions. Some suggestions for assisting such persons include:

- Their visual perception of written instructions or signs may be confused.
- Their sense of direction may be limited, requiring someone to accompany them.
- Directions or information may need to be broken down into simple steps. Be patient.
- Simple signals and/or symbols should be used (e.g., the graphics used throughout this section).
- A person's ability to understand speech is often more developed than his/her own vocabulary. Do not talk about a person to others in front of him/her.
- The individual should be treated as an adult who happens to have a cognitive or learning disability. Do not talk down to them or treat them as children.

Mobility Impairments

Someone using a crutch or a cane might be able to negotiate stairs independently. One hand is used to grasp the handrail the other hand is used for the crutch or cane. Here, it is best NOT to interfere with this person's movement. You might be of assistance by offering to carry the extra crutch. Also, if the stairs are crowded, you can act as a buffer and "run interference."

Wheelchair users are trained in special techniques to transfer from one chair to another. Depending on their upper body strength, they may he able to do much of the, work themselves. If you assist a wheelchair user, avoid putting pressure on the person's extremities and chest. Such pressure might cause spasms, pain and even restrict breathing. Carrying someone slung over your shoulders (something like the so called fireman's carry) is like sitting on their chest and poses danger for several individuals who fall within categories of neurologic and orthopedic disabilities.

One-Person Carry Technique

The CRADLE LIFT is the preferred method when the person to be carried has little or no arm strength. It is safer if the person being carried weighs less than the carrier's weight.

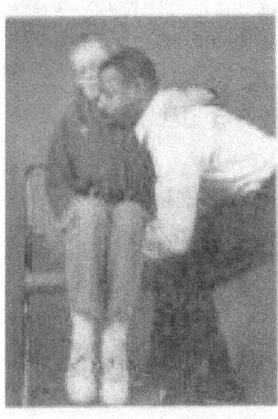

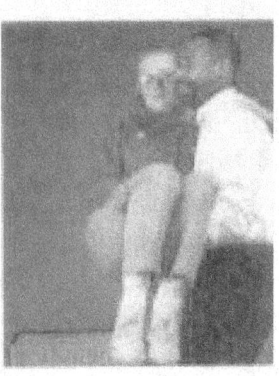

Two-Person Carry Technique — The Swing or Chair Carry

To use this technique

■ Carriers stand on opposite sides of the individual

■ Take the arm on your side and wrap it around your shoulder.

■ Grasp your carry partner's forearm behind the person in the small of the back.

■ Reach under the person's knees to grasp the wrist of your carry partner's other hand.

■ Both carry partners should then lean in, close to the person, and lift on the count of three.

■ Continue pressing into the person being carried for additional support in the carry.

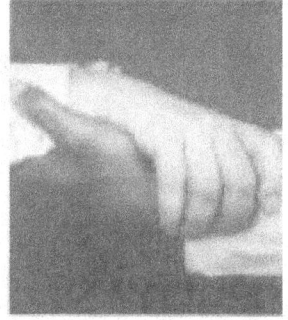

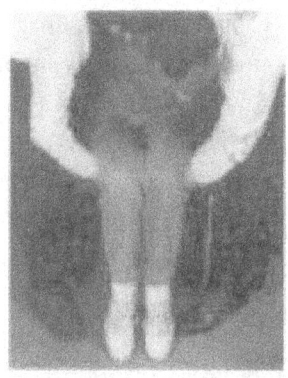

The advantage of this carry is that the partners can support (with practice and coordination) a person whose weight is same or even greater than their own weight.

The disadvantage is increased awkwardness in vertical travel (stair descent) due to the increased complexity of the two person carry. Three persons abreast may exceed the effective width of the stairway.

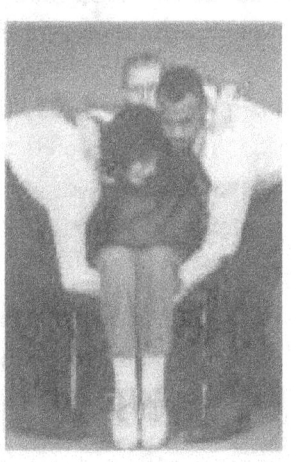

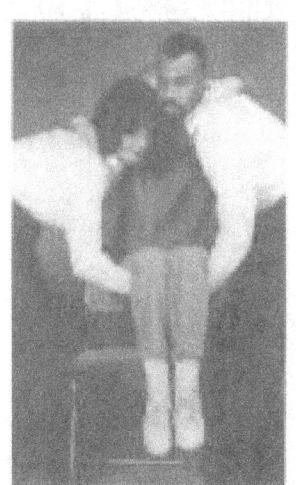

To Assist in Moving a Wheelchair Downstairs

When descending stairs, stand behind the chair grasping the pushing grips. Tilt the chair backwards until a balance is achieved. Descend frontward. Stand one step above the chair, keeping your center of gravity low and let the back wheels gradually lower to the next step. Be careful to keep the chair tilted back. If possible, have another person assist by holding the frame of the wheelchair and pushing in from the front. But do not lift the chair, as this places more weight on the individual behind.

Pregnancy is not usually considered a disability, but it can result in reduced stamina or impaired mobility, especially in negotiating stairs. In these cases, offer to walk with the woman and be of support both emotionally and physically. Remain with her until you have reached safety and she has a safe, warm place to sit,

With **respiratory disorders,** such as asthma or emphysema, the onset of symptoms can be triggered by stress, exertion, or exposure to small amounts of dust or smoke. Remind the individual to bring inhalation medication before leaving the work place.

Persons with **cardiac conditions,** should be reminded to take their medications. Offer them assistance in walking; they may have reduced stamina and require frequent rest periods.

It is vital to have a clear understanding through effective planning and practice with the local fire and rescue services regarding evacuation procedures for persons with disabilities. Opinions can vary among local fire departments, for example:

■ Whether individuals with disabilities should remain in their workplaces, assemble in an area of refuge to await the arrival of the fire fighters, or whether fellow workers should help with their immediate evacuation.

■ What evacuation techniques are to be used, in particular the carry techniques for getting non-ambulatory individuals down the stairs.

■ Whether dog guides should be permitted to evacuate down the stairway with their owners. There are examples of the fire department instructing that the dog be

After Working Hours

Most office fire fatalities occur outside of normal working hours. Here, fires can grow unnoticed and persons working alone can be cut off from their normal egress route. In many buildings, only a few people working late and the housekeeping staff are present at night. An employee with a mobility impairment who has relied on the elevator for access may need help to get down the stairs, but trained "buddies" are unavailable. To compensate, the individual should alert building security upon entering the building. Someone will then be ready to search for and assist the individual to safety, if needed. Alternatively, the person could be instructed to telephone the fire department as to their location when an emergency occurs.

Managers should ensure that shift workers and others who work on the premises outside normal hours, such as cleaners, are included. If there are employees whose knowledge of English may be limited, training should be given in a manner which they can understand. Non-English speakers and staff who have poor reading skills should be considered when written instructions are being prepared.

separated from its owner.

Whatever the plan, what is most

critical is that it be coordinated and practiced with the local fire and rescue service.

A

ADA Americans with Disabilities Act

ADAAG Americans with Disabilities Act Accessibility Guidelines

Aphasia Absence or impairment of the ability to communicate through speech, writing or signs

Area of Refuge Area of Rescue Assistance An area that has direct access to an exit, where people who are are unable to use stairs may remain temporarily in safety to await further instructions or assistance during emergency evacuation.

Asthma A complex disease which causes obstruction of the respiratory system

B

Back Pain Can be caused by congenital conditions, disease or injury, but for the millions of people affected by this condition the key element is always the same, pain.

Blind/Blindness Range of vision impairments from the inability to distinguish light and dark to a loss of part of the visual field or the inability to see detail. (see definition: Visually Impaired)

Brailler "Perkins Brailler," an all-purpose Braille writer.

Buddy System The system of assigning the appropriate individual(s) to assist in the evacuation of persons with disabilities.

C

Cane As used by the blind individual, the cane is a natural extension of the arm and hand and is used as an "information gathering" device (to locate familiar landmarks) for the purpose of establishing a clear path of travel. The conventional two-point touch system: The cane is moved from side to side in an arcing motion. The width of the arc is usually two inches to either side of the shoulders. As the cane touches to the left, the right foot should be forward. The cane tip touches in the opposite direction of the leading foot.

Chronic Obstructive Pulmonary Disease (COPD) Includes the diseases of chronic bronchitis and emphysema.

Cerebrovascular Accident (CVA) Localized brain damage due to a ruptured blood vessel in the brain; commonly called a stroke.

Cerebral Palsy Non-progressive disorder of the brain, results from damage to the nervous system at birth or in the first hours or days of life; not a disease.

Closed Circuit TV Magnifier (CCTV) Consists of television camera which takes the picture of the printed page and a television monitor which displays image in enlarged form.

Critical Language Terminology that is unacceptable and/or insulting to persons with disabilities. (see list on p. 24).

Deaf/Deafness Range of auditory impairments, from a total lack of sensitivity to sound to reduced sensitivity to certain sound frequencies.

Dog Guide Dog that has been specially trained to assist people who are blind, physically disabled or hearing-impaired.

Epilepsy Condition characterized by occasional seizures. A small fraction of those with epilepsy are

photosensitive. Seizures can be triggered by flashing lights.

Exercise Induced Bronchospasm (EIB) a form of asthma

F

Fingerspelling

When no sign language exists for a thought or concept, the word can be spelled out using the American Manual Alphabet.

G

Guide Dog

Proprietary name for a dog guide.

H

Handicapped

Critical language (see list on p. 24).

Hearing Impaired

Scale of hearing impairment ranges from mild hearing loss to profound deafness, the point at which the individual receives no benefit from aural input. Many hard-of-hearing persons are able to use residual hearing effectively with the assistance of

hearing aids (HA) or other sound-amplification devices, often augmented by lip reading. Hearing aids amplify background noises as well as voices, so noise caused by emergency conditions (alarm bells, people shouting, sirens, etc.) may rise to an uncomfortable level for the person with the hearing impairment.

Head Pointer Stick or rod which is attached to a person's head with a head band so that by moving the head, an individual can perform tasks that would ordinarily be performed by hand or finger movement.

Hemiplegia A disability resulting from a CVA which involves some degree of muscle weakness and motor skill loss on one side of the body.

I

Interpreter

Professional who assists a deaf person in communicating with hearing people who cannot sign.

L

Learning Disability An individual who may have difficulty recognizing or being motivated to act in an emergency. These individuals may also have difficulty in following anything other than a few simple instructions.

Little People

General term for persons of short stature who are less than 4' 10" or whose height is significantly below average.

Low Level Signage/Floor Proximity Exit Signs

are ususaly placed between 6" to 8" above the floor. A supplement to the required exit sign. The required exit signs are usually

located over the exits or near the ceiling, the first place to become obscured by smoke.

Low Vision can be moderate to severe vision impairment which includes difficulty in reading without magnification and seeing fine detail. Some persons with low vision may be considered legally blind.

M

Means of Egress

An accessible means of egress is one that complies with these following guidelines: a continuous and unobstructed way of exit travel from any point in a building or facility to a public way. A means of egress comprises vertical and horizontal travel and may include intervening room spaces, doorways, hallways, corridors, passageways, balconies ramps, stairs, enclosures, lobbies, horizontal exits, courts and yards. Areas of refuge or evacuation elevators MAY BE included as part of accessible means of egress. (Contact the authority having jurisdiction or refer to the building codes for the local application or definition.)

Mobility Impaired Employees with mobility impairments can vary in the degree of assistance that they require. The degree of impairment can range from walking with a slow gait (thus requiring more time to exit), to walking with a mobility aid such as a cane, crutches and/or braces, to the wheelchair user.

Monitors/Wardens Terms used to identify the different assignments made in the Occupant Emergency Plan. For example, the duties of the Monitor could

include assisting with the coordination of the evacuation of their floor or unit, identifying people with disabilities who require special assistance and coordinating assignment of "buddies" or assistants to stay with them (e.g., see excerpt from NYC *Fire Drill & Evacuation Rules for Offices* on p. 15).

Mouth Wand Rod with a tooth grip that is held in the mouth and used to perform tasks that would normally be performed by hand.

Normal Critical language (see list on p. 24).

Nystagmus Uncontrolled movement of the eye also noticeable by (characteristic) head tilt because it helps the person get a better focus on an object, thereby to see better.

Optical Character Reader Device that can be scanned over a printed page, reading the text aloud through a voice synthesis system. These may also have provision for reading directly from a computer disk containing a word processor file.

Opticon Device to enable a blind person to "read," consisting of a camera that converts print into an image of letters which are then produced via vibrations onto the finger.

Paraplegia Impairment or loss of motor and/or sensory function in the thoracic, lumbar or sacral segments of the spinal cord, affecting trunk and legs.

Post Polio Syndrome This affects individuals who have recovered from polio.

The symptoms include an increase in muscle weakness and an increase in respiratory weakness. Usually necessitates use of a wheelchair.

Quadriplegia (**see** tetraplegia).

"Seeing Eye" dog Proprietary name for dog guide.

Seizure Involuntary muscular contraction, a brief impairment or loss of consciousness, etc., resulting from a neurological condition, such as epilepsy.

Service Animal Trained dog or other animal that provides assistance to a person who is blind, deaf or mobility impaired. The animal can be identified by the presence of a harness or backpack.

Sign Language Means of communication used by persons who are deaf.

Speech Disorder
Limited or difficult speech patterns or without speech.

Tactile Signage
Signs or labels with Braille, raised letters or textured patterns that can be read tactilely by persons with visual impairments.

Tetraplegia
Impairment or loss of motor and/or sensory function in the cervical segments of the spinal cord, affecting arms, trunk and legs.

Text Telephone
Equipment that includes TTYs and employs interactive graphic communications through transmission of coded signals across the standard telephone network.

Victim Critical language (see list to the right).

Visually Impaired
A person with a vision impairment may be totally or legally blind. Legally blind implies that a person may be able to differentiate between light and dark or see very large objects, but may not be able to see anything clearly enough to depend on their vision in an emergency situation. This can also include persons with LOW VISION who can see well enough to walk but cannot read without magnification.

Wardens Persons assigned as coordinators of emergency actions by occupants of a single floor or part of a floor of a building.

Critical Language

Persons with disabilities are sensitive to the use of certain terms which are considered to be demeaning. When discussing evacuation plans with employees, the following terms should be avoided.

Ablebodied

Afflicted

Amputee

Cerebral Palsied

Confined to a wheelchair

Courageous

Crippled

Deaf and Dumb

Deaf/mute

Disease

Dwarf

Gimp

Handicapped

Normal

Patient

Physically challenged

Poor

Retard, retardate or retarded

Spastic

Suffering

Unfortunate

Victim

Wheelchair bound

U.S. Architectural & Transportation Barriers
Compliance Board
Lawrence W. Roffee, *Executive Director*
1331 F. Street, NW,
Washington, DC 20004- 1111
voice 1-8OO-USA-ABLE/872-2253
TTY 1-800-993-2822
Technical Assistance on Accessibility
Issues & the ADA

President's Committee on Employment of
People with Disabilities
Maggie Roffee, *Program Manager*
1331 "F" Street, NW
Washington, DC 20004
voice 202-376-6200
TTY 202-376-6205

Job Accommodation Network*
Anne E. Hirsh
West Virginia University
P.O. Box 6080
918 Chestnut Ridge Road, Suite 1
Morgantown, West Virginia 26506-6080
voice/TTY 1-800-526-7234

National Fire Protection Association
Learn Not To Burn Foundation
Sharon Gamache, *Executive Director*
P.O. Box 9101
Batterymarch Park
Quincy, Massachusetts 02269-9101
voice 617-770-3000
TTY 6 17-984-7880

National Electrical Manufacturers Association
Malcolm E. O'Hagan, *President*
2101 L Street, NW, Suite 300
Washington, DC 20037
voice 202-457-8400

Accessibility Equipment Manufacturers Assn.
Terry Nevins-Buchholtz, *Administrative Assistant*
P.O .Box 51784
2445 South Calhoun Road
New Berlin, Wisconsin 53151
voice 414-789-9890
* An affiliate of PCEPWD

Eastern Paralyzed Veterans Association
Buffalo Regional Office
Brian Black, *Assistant Director*
  of Building Code Standards
Buffalo Regional Office
111 West Huron Street
Buffalo, New York 14202
voice 716-856-6582

National Association of the Deaf
Nancy J. Bloch, *Executive Director*
814 Thayer Avenue
Silver Spring, Maryland 20910
voice (301) 587-1788
TTY (301) 587-1789

Self Help for Hard of Hearing People, Inc.
Donna L. Sorkin, *Executive Director*
7910 Woodmont Avenue, Suite 1200
Bethesda, Maryland 20814
voice 301-657-2248
TTY 301-657-2249

Telecommunications for the Deaf, Incorporated
Alfred Sonnenstrahl, *Executive Director*
8719 Colesville Rd, Suite 300
Silver Spring, Maryland 20910
voice 301-589-3786
TTY 301-589-3006

One example of reference material available:
1994 National Directory for TTYs with listings
of emergency numbers in the 50 states.

National Federation of the Blind
Patricia Maurer, *Coordinator, Community Relations*
1800 Johnson Street
Baltimore, Maryland 21230-4998
voice 410-659-9314

American Council of the Blind
Oral Miller, *National Representative*
1155 15th Street, NW, Suite 720
Washington, DC 20005
voice 202-467-5081 or 800-424-8666

page 4
Clarity Phone
three models for
business use: the
basic model, 2-line
phone & 2-line fea-
ture speakerphone.
Walker Equipment,
a Plantronics Com-
pany, P.O. Box 829,
Ringgold, GA 30736

page4
Intele-Modem
Ultractec™
450 Science Drive,
Madison, WI 53711,
voice/TDD 608-238-
5400.

page 5
Shake-up 9-V
smoke detector,
receiver & vibrator
from Weitbrecht
Communications
Devices for the Deaf,
2656 29th Street,
Suite 205, Santa
Monica, CA 90405,
voice/TDD 1-800-
233-9130.

page 6
Superprint ES™
Telecommunication
Device for the Deaf,
from Weitbrecht
Communications
Devices for the Deaf.

page 7
Office Case-
Alerting System:
(adapted from case
for hotel guest room)
from Weitbrecht
Communications
Devices for the Deaf.

page 7
Mult-Alert
SS 12/24ADA Series
signaling strobes
from System Sensor,
3825 Ohio Ave,
St.Charles, IL 60174,
voice
1-800-SENSOR2.

page 8
Talking Signs Inc.
8 12 North Blvd.,
Baton Rouge, LA
70802. *voice*
504-344-28 12.
Talking Signs@ were
developed by the
Smith-Kettlewell
Eye Research
Institute, Love
Electronics, Inc. and
Talking Signs, Inc.
Installation in New
York City, "Light
House for the
Blind," example of
working system in
the office building
setting.)

page 9
LABDATA™
page 5, issue Vol.20,
No. 1, 1990 a quar-
terly technical and
information publica-
tion by Underwriters
Laboratories Inc.,
333 Pfingsten Road,
Northbrook, IL 60062.

page 10
EVACU-TRAC™
Garaventa (Canada)
Ltd., 7505-134A
Street, Surrey, BC,
Canada V3W 7B3,
voice 1-800-663-6556

page 10
EVAC+CHAIR™
Corporation, I7 East
67 Street, New York,
NY 10021, voice
212-734-6222

page 11
LABDATA™ Vol.20,
No. 1, 1990, pages
12 & 14. A quarter-
ly technical and
information publica-
tion by Underwriters
Laboratories Inc.

page 11
EVAC-U-STRAPS,
were devised by a
wheelchair user in
Atlanta, GA.

page 11
SCALAMOBIL alber
TMI, Technomar-
keting, Inc, 307
Bacon Road, Rouge-
mont, NC 27572,
voice 919-477-1387

For further informa-
tion about accessi-
bility equipment
and manufacturers
contact: Accessibility
Equipment and
Manufacturers'
Association (AEMA),
P.O. Box 51784,
New Berlin, WI
5 3 15 1 , *voice/fax*
414-789-9900

page 13
TECH SHEET,
the diagrams are
taken from one of a
series of publications
on the design re-
quirements of the
ADA Accessibility
Guidelines (AADAG),
written and com-
piled by the Barrier
Free Environment,
Inc., design for peo-
ple of all ages and
all abilities, P.O. Box
30634, Raleigh, NC
27622.